NEXTGEN LEARNERS PRESENTS

Think Like a Brain!™ Collection

Luna Lefty & Roger Righty's Mixed Up Masterpiece

How Both Sides of Your Brain Work Together

BOOK 9

Think Like a Brain!™
Collection Overview

Think Like a Brain!™ is a wildly engaging, silly-smart children's book collection that invites kids to explore how their brains work from the inside out. Through playful characters, imaginative settings, and rhythm-rich storytelling, this collection teaches neuroscience, social-emotional learning (SEL), and self-understanding in ways children absorb without even realizing they are learning.

Instead of telling children how to behave or what to feel, Think Like a Brain!™ shows them why their brains respond the way they do. Each story transforms complex brain science into friendly characters and relatable moments, helping kids feel capable, curious, and empowered rather than confused or ashamed.

Designed for children first and embraced by parents and educators as preventative education, this collection supports:

- Emotional regulation and self-awareness
- Executive functioning skills
- Neurodivergent understanding and inclusion
- Brain-body connection and growth
- Compassion for self and others

With humor, heart, and factual accuracy at its core, Think Like a Brain!™ helps children build the skills many adults wish they had learned sooner, one playful page at a time.

One of the collection's most powerful features is how the characters become real-life reminders for children long after the story ends. Once kids learn who the characters are and what jobs they do, they can begin to recognize those same processes in themselves. A big feeling might become, "It looks like Amy is trying to tell you something," or a moment of pause might sound like, "What does your Fred think?" By naming experiences through familiar characters, children learn to notice what is happening inside their brains, reflect on it, and move forward more quickly with understanding rather than frustration.

Series 1:
The Brainy Bunch (Brain Anatomy)

Series Overview: Meet the brain's core parts as a friendly team. Kids learn what each part does and how they work together to help thinking, feeling, moving, and learning.

1. **The Brainy Bunch**
   Focus: Brain overview and teamwork
2. **Brainstem Benny & The Automatic Adventure**
   Focus: Breathing, heartbeat, survival basics
3. **Bella Cerebellum's Balance Ballet**
   Focus: Balance, coordination, practice
4. **Olly Occipital & The Lookout Tower**
   Focus: Vision, visual processing
5. **Tammy Temporal & The Soundy Sound Storm**
   Focus: Hearing, language, sound processing
6. **Perry the Parietal & The Body Map Bash**
   Focus: Touch, space, body awareness
7. **Hippocampus Hal & The Memory Maze**
   Focus: Memory, learning, recall
8. **Amy the Amygdala & The Overactive Alarm**
   Focus: Emotional alarm system
9. **Luna Lefty & Roger Righty's Mixed Up Masterpiece**
   Focus: Left/right hemispheres, teamwork
10. **Callie the Connector & The Brain Bridge**
    Focus: Corpus callosum, brain-to-brain communication

Series 2:
Cortex Castle (Frontal Lobe Executive Functions)

Series Overview: A lively castle of executive-function helpers. Kids practice planning, starting, stopping, focusing, flexible thinking, time skills, and self-talk, with Frontal Fred guiding thoughtful choices.

1. **Cortex Castle**
   Focus: Executive function overview
2. **Frontal Fred & The Decision Dilemma**
   Focus: Decision-making, prioritizing
3. **Impulsive Ian & The Button Box**
   Focus: Impulse control, pause
4. **Memory Molly & The Wandering Thoughts**
   Focus: Working memory, holding steps
5. **Focus Frannie & The Great Distraction Chase**
   Focus: Attention, refocusing
6. **Planella & The Checklist Parade**
   Focus: Planning, sequencing
7. **Flexi & The Changey-Changeroo Day**
   Focus: Cognitive flexibility
8. **Self-Talk Sally & The Thoughtful Echoes**
   Focus: Self-talk, emotional regulation
9. **Starter Stacy & Finisher Finn**
   Focus: Task initiation, task completion
10. M**ilo the Minute Keeper & Mae the Moment Seeker**
    Focus: Time awareness, pacing
11. **The Power of Pausing with Penny**
    Focus: Pause skills, regulation

Series 3:
Busy Brain Café (Neurodivergent Affirmations)

Series Overview: A welcoming café where different brain styles are celebrated. These stories affirm neurodivergent experiences while teaching self-understanding, communication, and compassion for others.

1. **Nathan Neuron & The Bouncy Brain**
   Focus: ADHD metaphor, movement, attention
2. **Sara Sprinkle the Sensa & The Super Sensitive Day**
   Focus: Sensory sensitivity, overload tools
3. **Overwhelmed Owen & The Senses Overload**
   Focus: Sensory overload, coping strategies
4. **Interoceptive Ivy & The Inside Signals**
   Focus: Interoception, body signals
5. **Zane's Zigzag Thoughts & The Path to Peace**
   Focus: Anxiety, grounding
6. **The Busy Brain Café**
   Focus: Neurodivergent inclusion, different processing styles
7. **The Listening Light**
   Focus: Nonverbal communication, understanding
8. **The Language of Love**
   Focus: Connection, communication, compassion
9. **Trying Theo & The How-To Hiccup**
   Focus: Persistence, frustration tolerance
10. **Drowsy Drew & The Nighttime Repair**
    Focus: Sleep, brain repair
11. **Burned-Out Bonnie & The Rebuild**
    Focus: Burnout, rest, recovery

Series 4:
Steady Systems Squad (Regulation & Growth)

Series Overview: A squad that explains how brains regulate, reset, rest, and grow stronger over time, including stress waves, sleep crews, sensory filtering, and rebuilding after hard days.

1. **The Steady Systems Squad**
   Focus: Regulation systems overview
2. **Timothy Thalamus & The Traffic Tower**
   Focus: Sensory filtering, attention routing
3. **Danny Dimmer & Wake-Up Wonda**
   Focus: Arousal and alertness
4. **The Start, Stop, and Switch Station**
   Focus: Initiation, inhibition, shifting
5. **The Night Shift & The Sleepy Crew**
   Focus: Sleep systems, restoration
6. **The Stress Wave & The Return to Calm**
   Focus: Stress response, recovery
7. **The Brain Growth Gathering**
   Focus: Brain growth and development
8. **Paul's Practice Path**
   Focus: Neuroplasticity
9. **Daisy's Daydream Network**
   Focus: Daydream network, imagination
10. **Remy's Rebuild Crew**
    Focus: Recovery and healing

NEXTGEN LEARNERS PRESENTS

Think Like a Brain!™ Collection

Luna Lefty & Roger Righty's Mixed Up Masterpiece

How Both Sides of Your Brain Work Together

BOOK 9

WRITTEN &
ILLUSTRATED BY:

MAISEL MCLAULA

NEXTGEN LEARNERS PRESS™
CINCINNATI, OHIO

This book is published by NextGen Learners Press, an imprint of NextGen Learners LLC.

Maisel McLaula is a pen name used by the author of this work.

First Edition: 2026
Printed in the United States of America
ISBN: 978-1-972860-08-3

This book is intended for educational and informational purposes only. It is not intended to replace professional medical, psychological, or therapeutic advice, diagnosis, or treatment.
Always seek the advice of a qualified professional with any questions regarding a medical or mental health condition.

Artwork and visual elements in this book were created using a combination of original artwork, with enhanced original direction, digital illustration tools, and AI-assisted creative technologies.

A Note for My Readers:

Hi friend,
Your brain is incredible. It thinks, feels, remembers, solves, dances, creates, and so much more!
This series is here to help you understand how your brain works, and to cheer you on as you grow your own thinking superpowers.

Every part of your brain has a job.
Sometimes they get mixed up. That's okay!

When you learn what each part does, you can work with your brain instead of feeling like it's working against you.
So read, giggle, learn, and maybe even draw your own Brainy Bunch when you're done.

I'm cheering you on with every rhyme.

-Maisel McLaula

Think Like a Brain!™

Brain Anatomy Collection

Explore the castle inside your head! Meet each room and its brilliant resident, from Frontal Fred to Benny the Brainstem. Whimsy, wonder, and neurons await!

For My Twins; Ale'na & Asa,
Two souls, forever connected, bound beyond time. One carries the world forward, the other lives in memory, together shaping logic and creativity as one. I love you beyond measure.

And...

To the beautifully wired, the curiously wired, and the wonderfully neurodivergent minds, both sides of you belong.

You are exactly what a great brain looks like.

-Maisel McLaula

Inside Cortex Castle, down a hallway so wide,
Twins lived on the brain's left and right side.
One named Luna, she loved logic and plans,
While Roger made pictures with his feet and his hands.

They shared the same tower, side by side each day,
But saw the world in a different way.

Luna loved patterns, steps, and rules,
Roger preferred colors, music, and tools.

One day they were given a marvelous task:
"To build a grand mural," said Fred with a laugh.

"It must be creative, but make perfect sense,
With color and structure and proper sequence!"

Luna Lefty made charts and lists in a row,
While Roger Righty just shouted, “Let’s go with the flow!”

He splashed on some glitter, a spiral, and sand,
While she panicked: “This wasn’t part of the plan!”

The mural was chaos, a jumble of bits,
Luna declared, "This gives me the fits!"

Roger just twirled and painted a bear,
As Luna re-labeled each section with care.

Then Fred reappeared with a knowing old grin,
"You need to stop fighting and try to begin,

To work side by side, not battle all day.
You each bring a strength in a different way."

So Luna laid out a structure with lines,
And Roger filled shapes with squiggles and signs.
Luna drew perfect petals, and Roger drew bees,
Together they balanced the flowers and trees.

The mural began to look rather grand,
With patterns and splashes from both of their hands.

It told a great story with beauty and grace,
And brought a huge smile to Fred's joyous face.

"You see," Fred explained, "your brain is one piece,
But two different halves help it work best in peace.

They handle different jobs in their own special way,
And work even better when both get to play."

Luna and Roger high-fived and then beamed,
Their teamwork was better than either had dreamed.

With logic and color, their mural stood proud,
And Cortex Castle cheered long and loud.

From that day on, they built side-by-side,
Designing and dreaming with nothing to hide.
For every good plan needs a spark to ignite,
And every wild thought needs a path that feels right.

So when your brain battles over what you should do,
Should you dance, or solve math, or paint something new?

Just know that both sides have something to say,
And they work best together when you let them both play!

What Are the Left and Right Hemispheres of the Brain?

Lefty's Strengths:

- Logic and order
- Math and patterns
- Language and steps
- Planning and rules

Righty's Strengths:

- Imagination and art
- Music and rhythm
- Emotions and play
- Seeing the world in a new kind of way

Try It Yourself!

- Draw your own Lefty and Righty mural
- Make one half full of patterns and plans
- And the other wild, colorful, and bright!
- Then blend the two in the middle.

Congratulations!

You've met your two halves and learned what they do.
One helps you think, and one dreams with you.
So treasure them both, and give them the stage,
Together they're brilliant, no matter your age!

Thank You to My Readers

Thank you for reading this book.
Whether you're a parent, a teacher,
a therapist, or a brilliant young
brain explorer,
your interest in understanding
how the mind works
means the world.

Sharing these books
means spreading compassion and
curiosity.

You are the reason
these stories come alive.
Thank you for believing in the
power of playful learning.

Stay curious. Stay kind.
Stay brainy!

Luna Lefty & Roger Righty's Mixed Up Masterpiece

Teacher & Parent Guide

💡 What This Story Teaches:

- Brain parts: Left and Right Hemispheres (thinking styles)

- Core idea: Different thinking styles work best together.

- Skills supported: problem-solving, creativity, logic, emotional expression, flexible thinking

- Big message: Your brain is strongest when both sides cooperate.

Luna and Roger show children that there is no "better" way to think. Each side brings something important to the team.

Key Concepts for Grown-Ups

Luna and Roger represent the two hemispheres of the brain. While both sides work together constantly, each tends to specialize in different styles of processing.

The left hemisphere often supports:

- Language and words
- Sequencing and steps
- Logic and details
- Math and structure

The right hemisphere often supports:

- Creativity and imagination
- Big-picture thinking
- Visual-spatial awareness
- Emotional tone and expression

Strong learning happens when information flows between both sides.

When integration is difficult, children may:

- Get stuck in rigid thinking
- Struggle to explain ideas
- Have trouble switching tasks
- Feel frustrated when logic and creativity clash
- Appear "one-sided" in their approach

Important foundations:

- Neither side works alone.
- Cross-body movement supports brain integration.
- Blending logic and creativity strengthens understanding.
- Flexible thinking grows with practice.

This story helps children understand that:

- Different brains think differently.
- Both styles are valuable.
- Teamwork inside the brain creates balance.

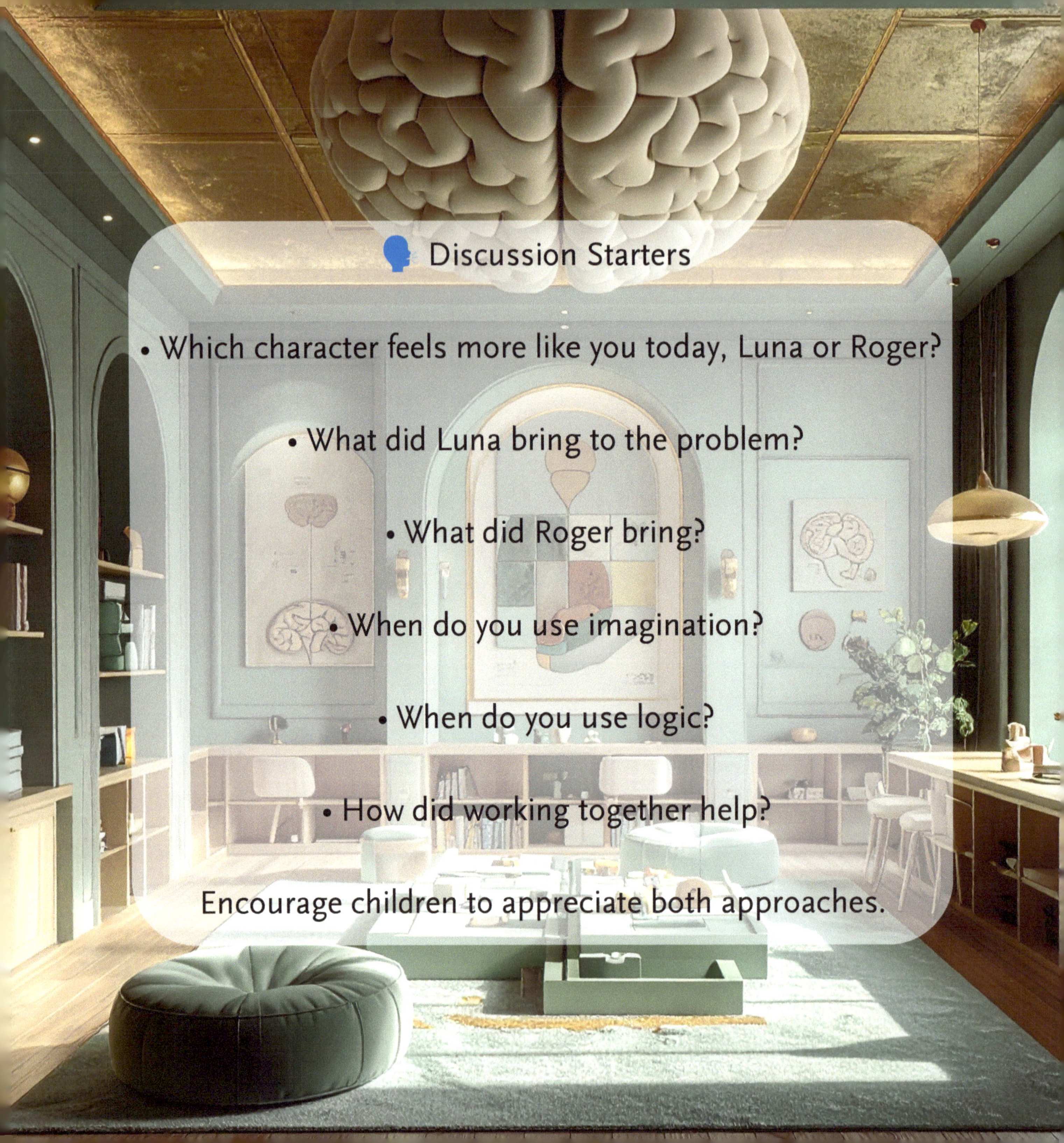

Discussion Starters

- Which character feels more like you today, Luna or Roger?
- What did Luna bring to the problem?
- What did Roger bring?
- When do you use imagination?
- When do you use logic?
- How did working together help?

Encourage children to appreciate both approaches.

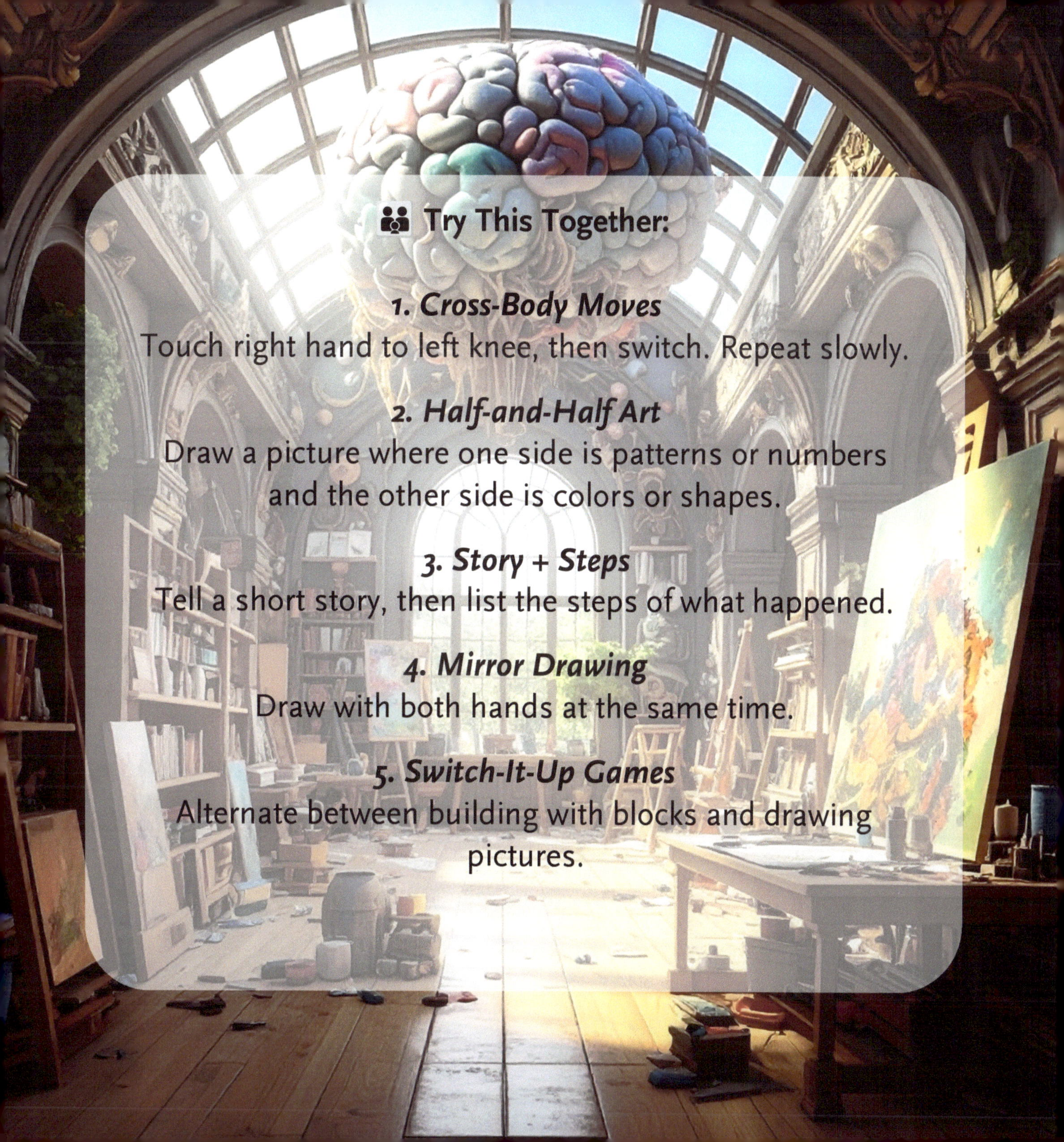

Try This Together:

1. Cross-Body Moves

Touch right hand to left knee, then switch. Repeat slowly.

2. Half-and-Half Art

Draw a picture where one side is patterns or numbers and the other side is colors or shapes.

3. Story + Steps

Tell a short story, then list the steps of what happened.

4. Mirror Drawing

Draw with both hands at the same time.

5. Switch-It-Up Games

Alternate between building with blocks and drawing pictures.

🎯 Bonus Challenge

Create Luna & Roger's Team Board

Divide a paper in half.

On one side, add creative ideas.

On the other side, add logical steps.

Ask:

- How did each side help?
- What happens when they work together?

Optional Labels:

- Imagination Zone
- Logic Lane
- Team Bridge

Playful Extension:

Brain Team Relay

Set up stations:

- Draw something silly
- Solve a simple puzzle
- Build a tower
- Tell a story

Rotate through each. Notice how different skills work together.

🛠️ Luna & Roger's Thinking Toolbox:

Use these tools when children struggle with flexible thinking:

- Encourage cross-body movement
- Blend art with academics
- Offer choices instead of commands
- Break problems into steps
- Invite imagination during learning
- Model switching strategies

Helpful phrases:

- "Let's use both sides."
- "What does Luna think?"
- "What would Roger try?"
- "Your brain is teaming up."

Avoid labeling children as only logical or only creative.

Integration builds resilience.

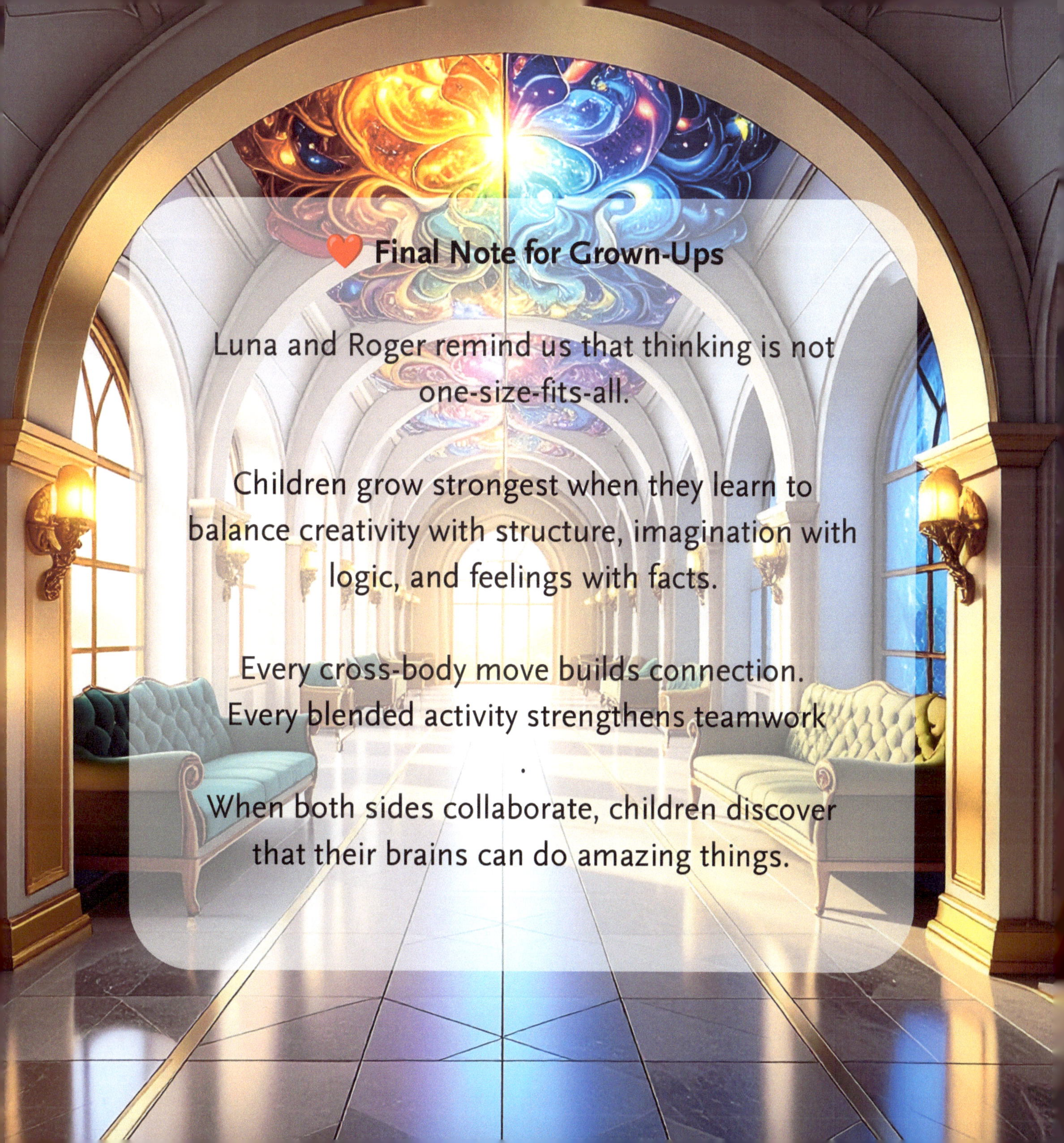

❤️ Final Note for Grown-Ups

Luna and Roger remind us that thinking is not one-size-fits-all.

Children grow strongest when they learn to balance creativity with structure, imagination with logic, and feelings with facts.

Every cross-body move builds connection.
Every blended activity strengthens teamwork
.
When both sides collaborate, children discover that their brains can do amazing things.

Resources & GROWTH

Resources:

(Left and right brain hemispheres + how the two sides work together)

- NINDS (NIH). Brain Basics: Know Your Brain (left/right hemispheres overview).
- Harvard Center on the Developing Child. InBrief: The Science of Early Childhood Development / brain architecture (for integration + development framing).
- NINDS (NIH). Brain Basics: Know Your Brain / Brain Basics resources.
- Cleveland Clinic. Brain: parts and function (overview pages).
- CASEL. CASEL's SEL Framework (self-awareness, self-management, social awareness, relationship skills, responsible decision-making).
- CDC. Child Development: Positive Parenting Tips and developmental guidance by age.
- Cleveland Clinic (left brain right brain overview)
- MedlinePlus (U.S. National Library of Medicine) (brain hemispheres)
- Encyclopedia Britannica (cerebral hemispheres)

Growth:

(If you or someone you know wants to strengthen whole-brain cooperation)

- Harvard Center on the Developing Child
  - Look for: integration, flexible thinking
- CASEL
  - Look for: social awareness

Acknowledgments:

To my students: past, present, and future:
You taught me more than any textbook ever could. By watching how you explored, questioned, struggled, and discovered, I learned how minds truly grow. Every "aha" moment in these pages began with you.

To the parents, teachers, and guides walking beside children every day:
Thank you for listening, for advocating when things feel complicated, and for believing that understanding a child's inner world truly matters. Your patience and care shape futures in ways you may never fully see.

To my husband, my best friend, and my constant partner in learning:
Thank you for walking beside me, believing in this work, and holding space for creativity to exist. Sharing curiosity, growth, and creation with you has been one of the greatest honors of my life.

To the great thinkers who shaped how I see the world: including Dr. Seuss, Albert Einstein, Leonardo da Vinci, Stephen Hawking, Ludwig van Beethoven, just to name a few, and my English teacher, Mrs. Godfrey:
Your curiosity, courage, and imagination reminded me that learning is limitless, creativity is powerful, and questions matter. Mrs. Godfrey, your encouragement, belief in me, and love of words sparked my writing journey. If not for you, I may never have picked up a pen.

And to my publisher, NextGen Learners™:
It was by fate that we met, and by inspiration that we now move forward. Together, we build something meaningful. Together, we leave this legacy.

Lastly, to my own brain:
Complex, resilient, ever-adapting, and endlessly learning. Thank you for carrying stories, holding wonder, and reminding me that growth is always possible.

Maisel McLaula is the author of the Think Like a Brain!™ children's book collection, a story-first series that blends neuroscience, imagination, and emotional safety for young readers.

Drawing from decades of experience supporting children and families, her work centers on a simple but powerful belief: children deserve to understand and accept who they are, including how their brains work and why they make the choices they do.

Through playful characters and gentle metaphors, Maisel's stories help children discover the "why" behind their thoughts, feelings, and actions, turning confusion into clarity and self-judgment into self-compassion. By understanding their inner systems, children gain the power to respond with kindness toward themselves and to recognize those same patterns in others.

Maisel believes that when children feel safe inside their own minds, empathy grows naturally. The Think Like a Brain!™ collection offers a foundation for confidence, understanding, and care that grows with the child.

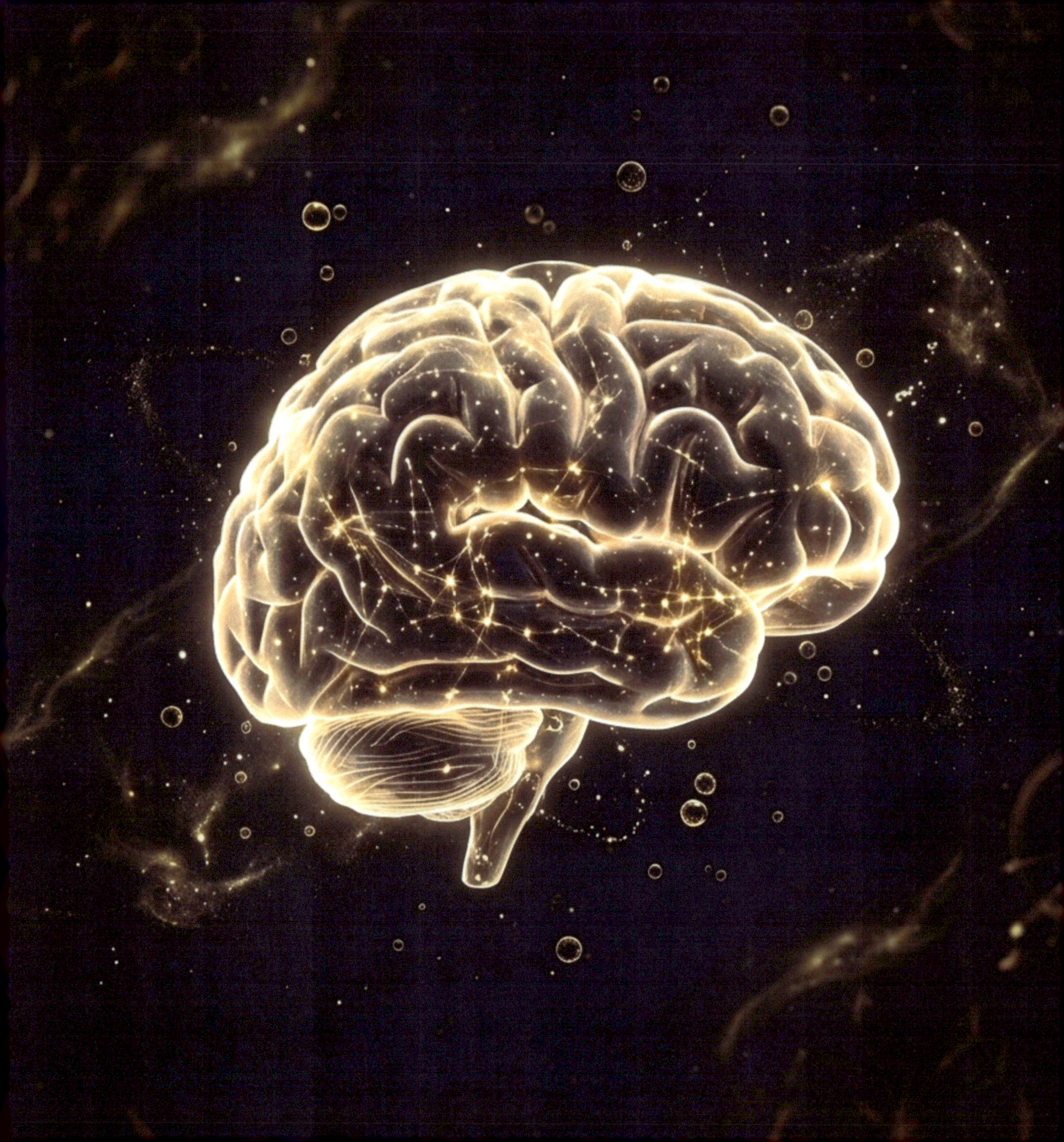

ABOUT THE PUBLISHER

NextGen Learners LLC is an educational publishing company dedicated to helping children understand how they think, feel, learn, and grow.

Through imaginative storytelling, neuroscience-informed concepts, and real-world application, NextGen Learners™ creates books and resources that support emotional development, critical thinking, self-awareness, and confidence in young minds.

Our work is designed not only for children, but for the parents, educators, caregivers, and professionals who walk beside them. Each book serves as a bridge, helping adults better understand the child in front of them, while helping children better understand themselves.

By creating shared language and meaningful connection points, our stories foster deeper relationships, reduce misunderstanding, and encourage acceptance, both within oneself and toward others.

We believe every child's mind is uniquely wired, and that understanding those differences is the foundation for compassion, growth, and lifelong learning.
With multiple collections designed to support cognitive, emotional, and developmental growth, NextGen Learners provides tools that children can carry with them, and tools that adults can use to guide them with clarity, patience, and purpose.

MORE TO EXPLORE

The Brainy Bunch is just the beginning.
More stories, tools, and brain-friendly resources are on the way! Designed to help children, parents, and educators better understand how the brain works in everyday life.

FREE PRINTABLES

We've created printable pages to go along with this book, including:

-Reading Guides

-Brain helper visuals

-Character Cards

These are available for free as part of our growing resource library.

Our full website experience is built to include:

-Downloadable activity packs

-Parent & teacher guides

-Expanded learning tools

-Additional collections and topics

STAY CONNECTED

We love hearing how children connect with these stories!

If there is a topic, feeling, or challenge you would like to see explored in future books, we would love to hear from you!

Contact us: Inspire@NextGenLearners.com
NextGenLearners.com

Your ideas help shape future stories and support the growth of this collection.

This collection grows with you. New resources and stories are added regularly.

Maisel McLaula loves hearing from the tiny humans who read her books.

Did a story make you giggle, spark a big idea, or inspire you to draw your very own character? She would be delighted to see it! Send your letters and artwork to:

NextGen Learners
Attn: Maisel McLaula
8666 Beechmont Ave #1063
Cincinnati, OH 45255

Every letter is read with a smile, and your artwork just might make her whole day.

OPEN BRAIN

www.ingramcontent.com/pod-product-compliance
Lightning Source LLC
Chambersburg PA
CBHW041322210926
51847CB00002B/26
9781972860083